Table of Contents

Please click on the Table of Contents entry below to be taken directly to a specific ship, center, aircraft, asset, program, or information. The fleet is listed based on the geographical location of their homeport/base starting in the Northeast and ending in the Pacific.

Office of Marine and Aviation Operations (OMAO)
and the NOAA Commissioned Officer Corps
- In the News -

Studying the Heart of El Niño, Where Its Weather Begins
-New York Times

HONOLULU — A thousand miles south of Hawaii, the air at 45,000 feet above the equatorial Pacific was a shimmering gumbo of thick storm clouds and icy cirrus haze, all cooked up by the overheated waters below. In a Gulfstream jet more accustomed to hunting hurricanes in the Atlantic, researchers with the National Oceanic and Atmospheric Administration were cruising this desolate stretch of tropical ocean where the northern and southern trade winds meet. It's an area that becalmed sailors have long called the doldrums, but this year it is anything but quiet. This is the heart of the strongest El Niño in a generation, one that is pumping moisture and energy into the atmosphere and, as a result, roiling weather worldwide...

NOAA Uses Improved UAS for Hurricane Analysis
-Flying

While manned airplanes have been used to analyze hurricanes there are times when conditions are simply too severe for safe flight. These are perfect opportunities to launch an unmanned aircraft system. The National Oceanic and Atmospheric Administration is using a light UAS called Coyote for the job. Coyote is launched from a tube attached to a Lockheed Martin P-3 Orion hurricane hunter and controlled remotely from the airplane. Unlike the P-3, the Coyote can linger inside the hurricane and drop down below 2,000 feet, which would be much too risky in the airplane. In this way, weather analysts aboard the P-3 and at NOAA's National Hurricane Center can gather data from the most severe parts of the storm through the UAS. Prior to the use of Coyote, NOAA would drop sensors out of the airplane to attempt to capture data...

A Global Hawk checks out El Nino
-WMDT

NASA and NOAA are collaborating to send NASA's Global Hawk out over the Pacific Ocean to monitor storms made by the strong El Nino we are experiencing. This is all part of the NOAA mission called Sensing Hazards with Operational Unmanned Technology (SHOUT)...Also as part of the SHOUT program, NOAA will also use a Gulfstream IV research plane and the NOAA Ship *Ronald H. Brown*. The goal is to get a clearer picture of El Nino and its impacts to West Coast U.S. storms and rainfall..

Unprecedented El Nino Study Uses Balloons, Aircraft
-NBC

Researchers launched weather balloons Tuesday off the coast of Hawaii in an unprecedented effort to discover how El Nino affects weather forecasts thousands of miles away. Craig McLean, assistant NOAA administrator for NOAA Research, explained how the project hopes to collect data from the Pacific Ocean using a research plane, a NOAA ship and drones...Eight times a day, weather balloons will be launched from the deck of the Ronald H. Brown as it sails from Honolulu to San Diego. It's expected to arrive here on March 18...The data collected by weather balloons like the ones launched Tuesday will be pulled along with data from instruments dropped from aircraft, researchers said...

OMAO Ships and Aircraft

– Mission Highlights –

NOAA Aircraft, Ship Support Major El Niño Study

NOAA aviators, mariners, and scientists are supporting a land, sea and air campaign in the tropical Pacific to study the current El Niño and gather data in an effort to improve weather forecasts thousands of miles away. The El Niño Rapid Response Field Campaign includes NOAA's Gulfstream IV aircraft, NOAA Ship *Ronald H. Brown*, NASA's Global Hawk unmanned aircraft equipped with NOAA sensors, and researchers stationed on Kiritimati (Christmas) Island in the Republic of Kiribati. "The rapid response field campaign will give us an unprecedented look at how the warm ocean is influencing the atmosphere at the heart of this very strong El Niño," said Craig McLean, assistant administrator for oceanic and atmospheric research.

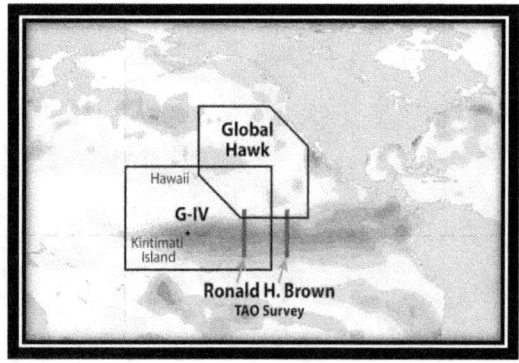

Graphic Illustrating operating area of NOAA assets during the El Niño Rapid Response Field Campaign.
[Graphic: NOAA/ESRL]

NOAA Hurricane Hunters Collect North Atlantic Ocean Winds Data

NOAA Lockheed WP-3D Orion N43RF departed its Tampa, Florida home base on Jan. 12 for Halifax, Nova Scotia, to support a NOAA Satellite and Information Service project, called "Ocean Winds," to validate satellite measurements of ocean surface wind speeds using equipment on board the aircraft as they fly through hurricane-force winter storms in the North Atlantic. These missions are carefully timed with satellite overpasses to provide valuable information that will aid in design of future observing systems and improve data sets for weather forecast models.

NOAA Flight Director Ian Sears aboard Lockheed WP-3D Orion N43RF during the Ocean Winds project.
[Graphic: NOAA/Terry Lynch]

NOAA Aircraft Respond to Midwest Floods

From Jan. 2-18, 2016, a team of NOAA aviators aboard NOAA's Beechcraft King Air 350 CER and Gulfstream Turbo (Jet Prop) Commander AC-695A collected aerial images of flooding along the Mississippi and Arkansas rivers using their specialized remote-sensing equipment. The imagery is available to the public online from the NOAA National Ocean Service's National Geodetic Survey. Natural resource and emergency managers rely on data collected by NOAA aircraft to inform decisions and create and update a variety of products, including damage assessments, flood and weather forecasts, and nautical charts.

NOAA Basic Officer Training Class (BOTC) 127

NOAA Officer Candidates have been very busy with training, drill competitions, and their Rules of the Road exam. Activities the students have taken part in include Helicopter Egress Training at Survival Systems USA, Inc., RADAR and ARPA training at Marine Simulation Institute, and a Drill Down competition against Coast Guard Academy cadets. The Drill Down competition gives NOAA Officer Candidates the opportunity to stand amongst their Coast Guard shipmates and represent Officer Candidate School (OCS) in a longstanding rivalry between OCS and the Corps of Cadets. This year, OCS made a clean sweep, holding all 4 positions in the final round.

Drill Down Competition: Officer Candidates compete with the Corps of Cadets in a display of military drill, bearing, and uniform wear.
[Photo: Petty Officer 2nd Class Richard Brahm, USCG]

Officer Candidates participating in Helicopter Egress Training at Survival Systems USA, Inc.
[Photo: Photo Credit: ENS DeCastro/NOAA]

Independent Review Team

In January 2016, OMAO convened an Independent Review Team (IRT) to conduct a review of our ship fleet and:

- The IRT will assess the health of the NOAA Fleet of research vessels, requirements for recapitalization, and analysis of operational, maintenance practices and technology infusion, as well as:
  - Utilization of alternatives to the NOAA Fleet (commercial contracting, Academic Research Fleet, other public-funded vessels) to meet requirements;
  - Analysis of current operational systems (crewing, scheduling) and current maintenance practices; and
  - Technology readiness and infusion (instrumentation and mechanical).
- The IRT will deliver a final report in September 2016.
- The guidance from the IRT will inform modernization plans of the NOAA fleet in alignment with the Federal Oceanographic fleet, charter options, as well as operation and maintenance strategies of the current fleet.
- The IRT process & report is independent of any larger Administration planning efforts related to the federal fleet.
- The IRT consists of twelve persons from across Federal government, Academia, and private sector to include expertise in science, ship-based data collection requirements, vessel operation, vessel design and building, and ship-based technological advancements

Independent Review Team

Co-Chairs:

Dick West
RADM, US Navy (ret)

Robert Winokur
Senior Advisor
Michigan Tech Research Institute / Michigan Tech Univ.

Members:

Fred Byus
RDML, US Navy (ret)
Vice President & General Manager
Battelle Mission and Defense Technologies

Dr. John Hughes-Clarke
Professor
University of New Hampshire

John Crowley
RADM, US Coast Guard (ret)
Executive Director
National Association of Waterfront Employers

Bauke (Bob) Houtman
Head, Integrative Programs Section
National Science Foundation, Ocean Sciences Division

Dr. Steve Murawski
Professor
University of South Florida

Blake Powell
President
JMS Naval Architects

Nancy Rabalais, Ph.D.
Executive Director and Professor
Louisiana Universities Marine Consortium
National Defense University, Penn State University

Dr. Steve Ramberg
Distinguished Research Fellow
Center for Technology and National Security Policy,

Robert (Tim) Schnoor
Ocean Research Facilities Manager
Office of Naval Research

Dick Vortmann
President and CEO (retired)
National Steel and Shipbuilding Company (NASSCO)

NOAA Liaisons:

Nancy Hann, CDR/NOAA
Chief of Staff
NOAA Office of Marine and Aviation Operations

Richard J. Park, LT/NOAA
Flag Aide to Director, NOAA Corps and OMAO
NOAA Office of Marine and Aviation Operations

OMAO's Ships and Centers

OMAO's Ship Tracker (screen shot below) shows information about the location - present and past - of our fleet of research and survey ships. Please note: To access Ship Tracker you must create an account with a **.gov** or **.mil** email address. All other access is restricted.

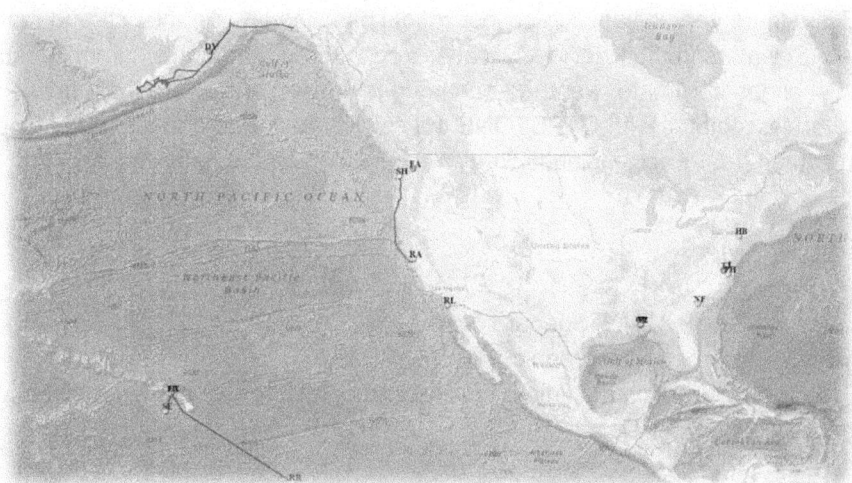

OMAO's ships and related Marine Centers are listed below based on the geographical location of the vessels' homeports starting in the Northeast and ending in the Pacific.

New Castle, NH
NOAA Ship *Ferdinand R. Hassler*

Commanding Officer:	LCDR Briana Welton
Primary Mission Category:	Hydrographic Surveys
DEPART: Norfolk, Virginia	**ARRIVE:** Norfolk, Virginia

Project: Approaches Chesapeake Bay
Objectives: To support safe navigation through the acquisition and processing of hydrographic survey data for updating nautical charts and by the identification and dissemination of dangers to navigation, as identified during the course of survey operations.

Woods Hole, MA (currently docks in Newport, RI)
NOAA Ship *Henry B. Bigelow*

Commanding Officer:	CDR G. Mark Miller
Primary Mission Category:	Fisheries Research
DEPART: Norfolk, Virginia	**ARRIVE:** Newport, Rhode Island

Project: Spring Bottom Trawl Survey
Objectives: Determine the spring distribution and relative abundance of fish and invertebrate species found on the continental shelf and upper slope, including the collection of additional biological information following the pre-established sampling plan at the direction of the Chief Scientist. Opportunistically evaluate survey gear efficiency, methods, and survey related equipment that may benefit the trawl survey and fish stock assessments. Collect oceanographic data including CTD casts and bongo tows at selected stations; and opportunistically collect acoustic data along cruise tracks with the EK-60 and ME-70 acoustic systems.

Davisville, RI
NOAA Ship *Okeanos Explorer*

Commanding Officer: CDR Mark Wetzler
Primary Mission Category: Oceanographic Exploration and Research
Depart: Honolulu, Hawaii **Arrive:** Kwajalein, Republic of the Marshall Islands
Depart: Kwajalein, RM **Arrive:** Apra Harbor, Guam

Project 1: CAPSTONE
Objectives: CAPSTONE is a three year initiative to collect critical baseline NOAA science and management needs in largely unknown areas of U.S. waters in the Pacific. Operations conducted during this campaign support NOAA missions to understand and predict changes in climate, weather, oceans and coasts, and share that knowledge and information with others. Much of this work associated with CAPSTONE will contribute to and complement Deep Sea Coral Research and Technology Program's three-year Pacific Islands Regional Initiative.

Norfolk, VA
NOAA Ship *Thomas Jefferson*

Commanding Officer: CAPT Shepard Smith
Primary Mission Category: Hydrographic Surveys
Ship Status: Alongside US Coast Guard Yard Curtis Bay - Baltimore, MD, for scheduled maintenance, winter repairs, scientific data processing, crew rest, and training.

OMAO'S MARINE OPERATIONS CENTER – ATLANTIC (MOC-A)
CAPT Anne Lynch, Commanding Officer MOC-A
MOC-A serves as a homeport for one NOAA ship, and manages the day-to-day operations and provides administrative, engineering, maintenance, and logistical support for the research and survey ships in NOAA's Atlantic fleet. Each year these ships conduct dozens of missions to assess fish and marine mammal stocks, conduct coral reef research, collect seafloor data to update nautical charts, and explore the ocean.

Charleston, SC
NOAA Ship *Nancy Foster*

Commanding Officer: LCDR Jeffrey Shoup
Primary Mission Category: Oceanographic Research, Environmental Assessment
Depart: Charleston, South Carolina **Arrive:** San Juan, Puerto Rico
Depart: San Juan, Puerto Rico **Arrive:** Ponce, Puerto Rico

Project: Essential Fish Habitat
Objectives: The Center for Coastal Monitoring and Assessment will be conducting the tenth year of an ongoing scientific research mission onboard NOAA Ship *Nancy Foster* funded by NOAA's Coral Reef Conservation Program. The purpose of the cruise will be to collect swath bathymetry, acoustical backscatter, ROV optical validation, fishery acoustics, and Slocum Glider deployments within coastal waters of Puerto Rico and St. Thomas, United States Virgin Islands.

Sunset from the NOAA Ship *Nancy Foster* at Detyens Shipyard in Charleston, SC
[Photo ENS Keith Hanson/NOAA]

NOAA Ship *Ronald H. Brown*

Commanding Officer: CAPT Robert Kamphaus
Primary Mission Category: Oceanographic Research, Environmental Assessment
DEPART: Honolulu, Hawaii   **ARRIVE:** San Diego, California

Project 1: TAO Maintenance (125°W and 140°W)
Objectives: Maintenance of the TAO moored ocean buoy array along the 125°W and 140°W meridians. The TAO buoy array is critical to providing real-time data for improved detection, understanding and prediction of El Nino and La Nina events.

Pascagoula, MS
NOAA Ship *Oregon II*

Commanding Officer: Master Dave Nelson
Primary Mission Category: Fisheries Research
DEPART: Pascagoula, Mississippi   **ARRIVE:** Galveston, Texas

Project 1: Experimental Longline Survey
Objectives: Conduct experimental bottom longline survey, on the U.S. continental shelf in the north east Gulf of Mexico. The comparison of bait types (squid vs. Atlantic Mackeral), gangion material (monofilament gangions vs. steel braided leaders), and deployment of a remotely operated vehicle for species identification and size; will be the primary focus of the investigation.

NOAA Ship *Pisces*
Commanding Officer: CAPT Michael Hopkins
Primary Mission Category: Fisheries Research
DEPART: Pascagoula, Mississippi   **ARRIVE:** Galveston, Texas

Project 1: SEAMAP Reef Fish Video Survey
Objectives: Conduct a survey of reef fish on the U.S. continental shelf of the GOM using a custom built stereo/video camera system and bandit reels. The ship's ME70 multibeam system and Simrad EK60 echosounder will be used to map predetermined targeted areas on a nightly basis to improve or increase the reef fish sample universe. The calibration of the EK60 and a patch test of the ME70 multibeam echosounder will be conducted when time and weather permit.

NOAA Ship *Gordon Gunter*
Commanding Officer: Master Donn Pratt
Primary Mission Category: Fisheries Research
DEPART: Pascagoula, Mississippi   **ARRIVE:** Pascagoula, Mississippi

Project 1: EPA Pascagoula Ocean Dredged Material Disposal Site Trend Assessment Survey
Objectives: The Pascagoula Ocean Dredged Material Disposal Site (ODMDS) is a large disposal site with an area of approximately 18.5 nautical miles. The site is located west of the Pascagoula Entrance Channel and due south of Horn Island. Historically, only the eastern portion nearest the channel has been utilized. However, there is a proposal to dispose of millions of cubic yards of dredged material from the expansion of the Gulfport turning basin into the western portion of the ODMDS. The ODMDS is used approximately every other year for disposal of maintenance material from the Civil Works Channel and Naval Station Pascagoula. San Diego, CA.

San Diego, CA
NOAA Ship *Reuben Lasker*
Commanding Officer: CDR John Crofts
Primary Mission Category: Fisheries Research
DEPART: San Diego, California   **ARRIVE:** San Francisco, California

Project 1: Coastal Pelagic Species
Objectives: Survey the distributions and abundances of pelagic fish stocks, their prey, and their biotic and abiotic environments in the area of the California Current between Newport, Oregon and Point Conception, California. The goal is to cover the northern inshore and offshore waters by occupying transect lines at 20 mile spacing in those areas where eggs and schools are present. If time allows, occupation of stations within the Southern California Bight will be conducted with data collected during the spring CalCOFI survey from the NOAA ship *Bell M. Shimada*. These stations will be added on an opportunistic basis.

Newport, OR
NOAA Ship *Rainier*
Commanding Officer: CDR E.J. Van Den Ameele
Primary Mission Category: Hydrographic Surveys
Ship Status: Alongside Newport, OR, for scheduled maintenance, winter repairs, scientific data processing, crew rest, and training and will be transiting to drydock in San Francisco, CA.

The NOAA Ship *Rainier* transits under the Golden Gate Bridge in San Francisco, CA.
[Photo: Lindsey Houska]

NOAA Ship *Bell M. Shimada*
Commanding Officer: CDR Paul Kunicki
Primary Mission Category: Fisheries Research
DEPART: Newport, Oregon **ARRIVE:** Newport, Oregon
DEPART: Newport, Oregon **ARRIVE:** San Diego, California
DEPART San Diego, California **ARRIVE:** San Francisco, California

Project 1: Southern Resident Killer Whale Critical Habitat Assessment
Objectives: Conduct acoustic and visual surveys of marine mammals and seabirds along the Oregon, Washington, California, and Canadian coasts in order to determine Critical Habitat in the coastal portion of the range of Southern Resident killer whales. The collection of predation, fecal, and biopsy samples will be of significant additional value. Secondary objectives include locating and documenting other cetacean species, in particular the collection of photographs and audio recordings of other killer whale pods, as well as sea bird counts and oceanographic data.

Project 2: CalCOFI
Objectives: Survey the distributions and abundances of pelagic fish stocks, their prey, and their biotic and abiotic environments in the area of the California Current between San Francisco, California and San Diego, California.

OMAO'S MARINE OPERATIONS
CAPT Todd Bridgeman, Director of Marine Operations
OMAO's Marine Operations over-sees operations of the three regional Centers, including the Marine Operations Center-Pacific, Marine Operations Center-Atlantic, and Marine Operations Center-Pacific Islands.

OMAO'S MARINE OPERATIONS CENTER – PACIFIC (MOC-P)
CDR Brian Parker, Commanding Officer MOC-P
MOC-P serves as a homeport for two NOAA ships, and manages the day-to-day operations and provides administrative, engineering, maintenance, and logistical support for the research and survey ships in NOAA's Pacific fleet. Each year these ships conduct dozens of missions to assess fish and marine mammal stocks, conduct coral reef research, collect seafloor data to update nautical charts, and explore the ocean.

Ketchikan, AK
NOAA Ship *Fairweather*
Commanding Officer: CDR David Zezula
Primary Mission Category: Hydrographic Surveys
Ship Status: Alongside Seattle, WA, for scheduled maintenance, winter repairs, scientific data processing, crew rest, and training.

Kodiak, AK
NOAA Ship *Oscar Dyson*
Commanding Officer: CDR Arthur "Jesse" Stark
Primary Mission Category: Fisheries Research
Depart: Newport, Ore.   **Arrive:** Kodiak, Alaska
Depart: Kodiak, Alaska   **Arrive:** Dutch Harbor, Alaska

Project 1: Acoustic-trawl survey of Shumagin Islands, Sanak trough, Pavlof Bay and Morzhovoi Bay.
Objective: Collect acoustic-trawl data necessary to determine the distribution, biomass, and biological composition of walleye Pollock; collect target strength data using hull-mounted transducers for use in scaling acoustic data to estimates of absolute abundance; collect physical oceanographic data (temperature and salinity profiles) at selected sites, and continuously collect sea surface temperature and salinity data; and conduct trawl hauls to ground truth multi-frequency echo integration data collection.

Project 2: Ecology of Ice-associated Seals in the Bering Sea
Objective: Study of the habitat requirements and ecological relationships with sea ice, of ribbon and spotted seals in the core of their Bering Sea breeding area. The seals' movements, haul-out behavior, diet, genetic population structure, health will be investigated and monitored. A particular focus of the work in 2016 will be on health and condition of young-of-the-year seals and potential impacts of diminishing sea ice.

Honolulu, HI
NOAA Ship *Hi'ialakai*
Commanding Officer: CDR Elizabeth Kretovic
Primary Mission Category: Oceanographic Research, Environmental Assessment
Ship Status: Alongside Honolulu, HI, for scheduled maintenance, winter repairs, scientific data processing, crew rest, and training.

NOAA Ship *Oscar Elton Sette*
Commanding Officer: LCDR Keith Golden
Primary Mission Category: Fisheries Research
DEPART: Pago Pago, American Samoa **ARRIVE:** Apia, American Samoa
DEPART: Apia, American Samoa   **ARRIVE:** Apia, American Samoa

Project: Samoan Archipelago Fisheries Life History
Objectives: Support deep-slope and shallow-water bottom fish, coastal pelagic fishes, and coral reef fishes bio sampling, collection of larval and juvenile stage pelagic and bottom fish species, surveys of coral reef ecosystems, and exploration of seamount benthic species, through 1) collection of adult deep-slope bottom fish, coral reef fish, and coastal pelagic fishes, 2) collection of pelagic stage (larvae and juveniles) eteline snappers and tuna, 3) daylight coral reefs snorkel surveys, 4) collection of oceanographic data from routine conductivity, temperature, depth casts, continuous acoustic doppler current profiler, and thermosalinograph measurements and, 5) collection of fishes and invertebrates at offshore seamounts using strings of Fathoms Plus traps.

OMAO'S MARINE OPERATIONS CENTER – PACIFIC ISLANDS (MOC-PI)
CDR Matthew Wingate, Commanding Officer MOC-PI
MOC-PI serves as a homeport for two NOAA ships, and manages the day-to-day operations and provides administrative, engineering, maintenance, and logistical support for the ships in NOAA's Pacific Islands' fleet.

OMAO's Aircraft

Tampa, Florida
WP-3D (N42RF) – "Hurricane Hunter"

Aircraft Commander:	N/A
Temporary Base:	Naval Air Station Jacksonville, FL
Current Mission:	Scheduled Maintenance - Until June 2016

The aircraft is at the Naval Air Station Jacksonville, Florida undergoing an extensive refurbishment period which will include replacing the wings and upgrading various components. This effort will extend the useful life of the aircraft for another 15-20 years.

UPDATE - Delay in WP-3D Service Life Extension Timetable

- The first NOAA P-3 (N42RF) currently in maintenance at *the U.S. Navy Fleet Readiness Center Southeast (FRCSE)* in Jacksonville, FL, will experience a 1 month delay in completion of its service life extension. This delay will have no scheduled impacts on NOAA programs, and the aircraft is scheduled to return to service for the last four months of the 2016 hurricane season.
- NOAA's second P-3 (N43RF) was scheduled to enter its service life extension maintenance period in Aug 2016. This project has experienced a 6 month delay due to an increased workload at FRCSE. The updated timeline now calls for the service life extension to begin in Feb 2017 and be completed approximately 14 months later.
- ***There are no programmatic impacts of the 1 month delay for N42RF or the 6 month slip for N43RF***. Due to the schedule changes, NOAA will now have 2 aircraft available for the 2016 Hurricane season and one available for the 2017 Hurricane Season, the scenario before the schedule changes.

WP-3D (N43RF) – "Hurricane Hunter"

Aircraft Commander:	CDR Price
Temporary Base:	Alaska and Greenland
Current Mission:	NASA Ice Bridge

NASA's Operation IceBridge images Earth's polar ice in unprecedented detail to better understand processes that connect the Polar Regions with the global climate system. Utilizing NOAA's highly specialized research aircraft, IceBridge employs the most sophisticated suite of science instruments ever assembled to characterize annual changes in thickness of sea ice, glaciers, and ice sheets. In addition, IceBridge collects critical data used to predict the response of Earth's polar ice to climate change and resulting sea-level rise. IceBridge also helps bridge the gap in polar observations between NASA's ICESat satellite missions.

Jet Prop Commander (N45RF)

Aircraft Commander:	LT Salling/ LTJG Doremus
Temporary Base:	Various locations
Current Mission:	Soil Moisture Surveys

NOAA aircraft use specialized detection equipment to make accurate, real-time measurements of snowpack characteristics and soil moisture across the country. This information is critical for managers and others to make optimal decisions supporting river, flood, and water supply forecasting, agriculture and forest management, recreation and winter tourism, and the commerce, industry, and transportation sectors of the Nation's economy. A single snowmelt flood can cause billions of dollars in damage and in the western areas of the country spring snowmelt provides over 70% of the annual water supply. The benefits of accurate snow and soil moisture measurements are immense and NOAA aircraft are uniquely capable to provide this information.

Gulfstream IV (N49RF)

Aircraft Commander: N/A

Current Mission: Maintenance (through April 2016)

The aircraft is undergoing scheduled maintenance for inspections and equipment upgrades that will increase capability to support NOAA science and extend the life of the aircraft.

Twin Otter (N46RF)

Aircraft Commander: LT Evory/LT Norman

Temporary Base: Various locations

Current Mission: Soil Moisture Surveys

NOAA aircraft use specialized detection equipment to make accurate, real-time measurements of snowpack characteristics and soil moisture across the country. This information is critical for managers and others to make optimal decisions supporting river, flood, and water supply forecasting, agriculture and forest management, recreation and winter tourism, and the commerce, industry, and transportation sectors of the Nation's economy. A single snowmelt flood can cause billions of dollars in damage and in the western areas of the country spring snowmelt provides over 70% of the annual water supply. The benefits of accurate snow and soil moisture measurements are immense and NOAA aircraft are uniquely capable to provide this information.

Twin Otter (N48RF)

Aircraft Commander: N/A

Current Mission: Maintenance (through March 2016)

The aircraft is undergoing scheduled maintenance for inspections and equipment upgrades that will increase capability to support NOAA science and extend the life of the aircraft.

Twin Otter (N56RF)

Aircraft Commander: LT Marino/ENS Blaauboer

Temporary base: East Coast, US

Current Mission: North Atlantic Right Whale surveys

North Atlantic right whales are critically endangered and listed under the Marine Mammal Protection Act. Aerial surveys serve multiple objectives with regard to conservation including providing locations and distribution of right whales to mariners to avoid collisions with ships, photo identification records on right whales, information on distribution and abundance of marine mammals and turtles, and provide sightings of dead whales for monitoring mortality.

Twin Otter (N57RF)

Aircraft Commander: LTJG Coker/LTJG Hirsch

Temporary base: Various Locations

Current Mission: TopoBathy LiDAR

The TopoBathy LIDAR mission will collect data in the coastal zone used to produce the most up-to-date- and accurate marine navigation charts, FEMA flood plain and inundation maps, and other Integrated Ocean and Coastal Mapping (IOCM) applications. Data gathered will help ensure safe and efficient marine transportation and benefit coastal communities with accurate resource management and aid emergency response efforts.

King Air (N68RF)

Aircraft Commander: LCDR Waddington/LT Sims

Current Mission: Various Locations – Continuous Coastal Mapping

Coastal Mapping is an on-going mission of NOAA's National Geodetic Survey (NGS) to survey approximately 95,000 miles of United States coastline providing the Nation with an accurate, up-to-date and seamless database of the national shoreline. This data is used as the baseline for defining America's marine territorial limits, including its Exclusive Economic Zone, and for the geographic reference needed to manage coastal resources and support marine navigation. Stereo photogrammetry and Light Detection and Ranging (LiDAR) are used to produce a digital database. In addition, the Coastal Mapping Program supports NOAA's homeland security and emergency response requirements by rapidly acquiring and disseminating a variety of datasets to federal, state, and local government agencies as well as the general public

OMAO'S AIRCRAFT OPERATIONS CENTER (AOC)
CAPT Michael Silah, Commanding Officer AOC

The AOC, located at MacDill Air Force Base in Tampa, Florida, serves as the main base for OMAO's fleet of nine aircraft and provides capable, mission-ready aircraft and professional crews to the scientific community. Whether studying global climate change or acid rain, assessing marine mammal populations, surveying coastal erosion, investigating oil spills, flight checking aeronautical charts, or improving hurricane prediction models, the AOC flight crews continue to operate in some of the world's most demanding flight regimes.

NOAA's Gulfstream-IV and its crew set out from Honolulu continue to fly a series of important missions in support of the NOAA Office of Oceanic and Atmospheric Research El Niño Rapid Response Field Campaign.
[Photo: NOAA]

Unmanned Systems Support

NASA Global Hawk
Location: Edwards Air Force Base (AFB), CA/ NASA Wallops Flight facility
Mission: Maintenance

NASA's Global Hawk Unmanned Aircraft System (UAS) successfully completed the NOAA-funded El Niño Southern Oscillation Rapid Response series of flights and will undergo required maintenance in March.

APH-22 Hexacopter
Location: Everett, Washington
Mission: Levee Setback Environmental Condition Monitoring

The NOAA Northwest Restoration Center (NWRC) seeks to add another layer of information to the monitoring effort on levees in the Snohomish River estuary in Puget Sound by utilizing data collected by the APH-22 UAS. The broad goals of the project are to transform the site into a vegetated, self-sustaining wetland that will 1) maximize the modern, natural ecological potential of the site 2) minimize adverse effects on, and add socio-economic value to the surrounding community and 3) advance the science and practice of restoration. Monitoring is critical in realizing and evaluating the performance of the project.

Location: Cape Shirreff, Livingston Island, Antarctica
Mission: Antarctica Marine Mammal Project

NOAA's Southwest Fisheries Science Center is deploying the APH-22 UAS platform for penguin and fur seal surveys on Cape Shirreff. This season's efforts will focus on utilizing the UAS for collection of replicate counts of breeding pairs and chicks for Gentoo and chinstrap penguins, Antarctic fur seal pup counts and defining the relationship between mass of leopard seals and their size and shape as determined from vertical aerial photographs. This later goal is especially important because the other alternative is to drug and capture the animals, which can be dangerous for both the scientist and the animals studied. In addition to these goals, studies on wildlife response to UAVs will be conducted with Antarctic fur seals, elephant seals, chinstrap, and Gentoo penguins.

Location: Cape Cod, MA
Mission: Cape Cod Whale Photogrammetry

Vessel surveys will be conducted in Cape Cod Bay during the spring feeding season for North Atlantic Right whales. The surveys will be conducted on most good weather days during the time the whales inhabit the Cape Cod Bay waters during March, April and May 2016. All surveys will be in coordination with other agency vessels and a survey plane. The APH-22, operated by 2 NOAA pilots-in-command (PICs), will be used to obtain aerial images of the whales. Photogrammetry will be employed to acquire accurate length measurements and length to width ratios of all demographics of right whales. Length measurements of calves will also be used to compare previous measurements taken in the calving grounds to estimate calf growth rates.

Puma UAS
Location: USCGC Polar Star (WAGB10)
Mission: VORTEX Convective Initiation

The NOAA VORTEX SE program plans to measure the conditions that lead to Convective Initiation (CI) in the lower boundary layer in Northern Alabama. A PUMA UAS system (N542FC), owned by NOAA/ATDD, will be used to measure the dynamics of land-atmosphere interactions in the lower boundary layer. ATDD's DJI S-1000 will also be used to perform storm damage assessment over a large area of Northern Alabama. The visible and near infrared cameras installed on the S-1000 will be used to document storm damage to assist the National Weather Service with determining the category of any tornado activity in the area that occurs during the VORTEX SE intensive study periods

OMAO Partnerships

United States Senate Committee on Commerce, Science, and Transportation

Location: Washington, DC

Detail: LCDR Wendy Lewis, NOAA Commissioned Officer Corps

LCDR Lewis is currently on detail to the Committee with the staff of the Chair, Senator John Thune (R-SD), where she is assisting on activities pertaining to oceans, atmosphere, and fisheries policy, as well as other matters within the Committee's jurisdiction.

National Science Foundation

Location: Antarctica

Mission: LT Jesse Milton, NOAA Commissioned Officer Corps

Members of the NOAA Commissioned Officer Corps carry out NOAA's mission in remote locations across the globe. LT Milton is assigned to Antarctica where he serves as the Station Chief for NOAA's Atmospheric Research Observatory (ARO) at the Amundsen-Scott South Pole Station. The ARO at the Amundsen-Scott South Pole Station is a National Science Foundation facility used in support of scientific research related to atmospheric phenomena.

Department of Defense - U.S. Pacific Command (USPACOM)

Location: Honolulu, Hawaii

Embedded Liaison: CAPT Barry Choy, NOAA Commissioned Officer Corps

The U.S. Pacific Command (USPACOM) area of responsibility encompasses approximately half the earth's surface and more than half of its population. The 36 nations that comprise the Asia-Pacific include: two of the three largest economies and nine of the ten smallest; the most populous nation; the largest democracy; the largest Muslim-majority nation; and the smallest republic in the world. The region is a vital driver of the global economy and includes the world's busiest international sea lanes and nine of the ten largest ports. By any meaningful measure, the Asia-Pacific is also the most militarized region in the world, with seven of the world's ten largest standing militaries and five of the world's declared nuclear nations. Under these circumstances, the strategic complexity facing the region is unique. CAPT Choy is linked closely with the activities within the region allowing for identification of opportunities and cooperation between USPACOM and NOAA, and better overall government function situational awareness in the region.

Department of Defense - U.S. Northern Command (USNORTHCOM)

Location: Boulder, Colorado

Embedded Liaison: CAPT Mark Moran, NOAA Commissioned Officer Corps

U.S. Northern Command (USNORTHCOM) partners to conduct homeland defense, civil support, and security cooperation to defend and secure the United States and its interests. NORTHCOM's area of responsibility includes air, land, and sea approaches and encompasses the continental United States, Alaska, Canada, Mexico, and the surrounding water out to approximately 500 nautical miles. It also includes the Gulf of Mexico, the Straits of Florida, and portions of the Caribbean region that include The Bahamas, Puerto Rico, and the U.S. Virgin Islands. CAPT Moran serves as the liaison for the NOAA Corps, helping to plan, organize, and execute homeland defense and civil support missions.

Department of Defense - U.S. Navy

Location: Washington, DC
Embedded Liaison: CDR Christiaan van Westendorp, NOAA Commissioned Officer Corps

CDR van Westendorp serves as NOAA liaison to the Oceanographer of the Navy and is an important interface between the U.S. Navy and other U.S. Federal Agencies, including NOAA. As NOAA Liaison, CDR van Westendorp serves as the Head of the Interagency Policy Branch of the International and Interagency Policy Division, Office of the Oceanographer of the Navy, located at the U.S. Naval Observatory. The mission of this Division is to coordinate and execute the Oceanographer of the Navy functions related to policy and programs involving international and/or interagency oceanography. Oceanography includes meteorology, oceanography, mapping, charting and geodesy, astronomy, and precise time and time interval.

Location: Stennis Space Center, Mississippi
Embedded Liaison: LTJG Laura Dwyer, NOAA Commissioned Officer Corps

Embedded in the Navy's Naval Oceanography Mine Warfare Center, LTJG Laura Dwyer works side by side with Navy officers operating Unmanned Underwater Vehicles worldwide and is currently deployed to the Arabian Gulf. This collaboration will provide knowledge and experience that will keep NOAA on the cutting edge of this emerging technology as well as strengthen the partnership between NOAA and the Navy.

Department of Homeland Security - U.S. Coast Guard

Location: Washington, DC
Embedded Liaison: CAPT Scott Sirois, NOAA Commissioned Officer Corps

As the NOAA liaison to the United States Coast Guard (USCG), CAPT Sirois maintains a current and comprehensive knowledge of interagency activities and policies related to the USCG and NOAA. He identifies potential conflicts or benefits issues for analysis and evaluation, conducts appropriate assessments and studies, and serves as the interface between NOAA and the USCG. CAPT Sirois initiates, designs, and implements strategies through federal agency liaison and coordination that results in cooperative arrangements for maritime security, oceanographic research, hazardous materials spill response, and many other activities.

Consortium for Ocean Leadership

Location: Washington, DC
Embedded Liaison: LCDR Josh Slater, NOAA Commissioned Officer Corps

LCDR Josh Slater serves as the NOAA liaison to the Consortium for Ocean Leadership (COL) and maintains a current and comprehensive knowledge of activities and policies related to COL's work and NOAA. The Consortium for Ocean Leadership represents more than 100 of the leading public and private ocean research and education institutions, aquaria and industry with the mission to advance research, education and sound ocean policy.

Teacher At Sea Program

The mission of the Teacher at Sea (TAS) program is to give teachers a clearer insight into our ocean planet, a greater understanding of maritime work and studies, and to increase their level of environmental literacy by fostering an interdisciplinary research experience. The program provides a unique environment for learning and teaching by sending kindergarten through college-level teachers to sea aboard NOAA research and survey ships to work under the tutelage of scientists and crew. Then, armed with new understanding and experience, teachers bring this knowledge back to their classrooms. Since its inception in 1990, the program has enabled more than 600 teachers to gain first-hand experience of science and life at sea. By participating in this program, teachers enrich their classroom curricula with knowledge that can only be gained by living and working side-by-side, day and night, with those who contribute to the world's body of oceanic and atmospheric scientific knowledge. Below is a list of the NOAA Teachers at Sea for the current monthly update for the 2015 Field Season. Once they have embarked on their cruise, you can gain access to their blogs which document their missions at sea and offer a wealth of information about the research being conducted as well as personal stories.

2015 Season Stats: 21 teachers sailed on different projects on NOAA vessels

2015 TAS Placements Blogs

Teacher-At-Sea, DJ Kast prepares to launch a drifter buoy as part of NOAA's Adopt-a-Drifter Program.
[Photo: NOAA]

OMAO - NOAA Dive Program

OMAO manages and implements NOAA's Dive Program (NDP), which trains and certifies scientists, engineers, and technicians from federal, state, tribal governments, and the private sector to perform the variety of tasks carried out underwater to support NOAA's mission. NDP also has cooperative diving agreements with over 100 government agencies and academic institutions. NOAA has more than 400 divers who perform over 14,000 dives per year. The NDP is headquartered at the NOAA Diving Center at the NOAA Western Regional Center in Seattle, Washington.

Trained NOAA divers help remove marine debris including these ghost fishing nets in Hawaii.
[Photo: NOAA]

OMAO Small Boat Program

OMAO manages NOAA's Small Boat Program and sets policy and provides safety inspections for almost 400 small boats operated by the various Line and program offices throughout NOAA, which support fisheries laboratories, dive support, nautical charting, ocean and Great Lakes research, and more.

NOAA small boats support many diverse operations across the country.
[Photos: NOAA]

Office of Marine and Aviation Operations

Providing environmental intelligence for a dynamic world

The personnel, ships, and aircraft of NOAA play a critical role in gathering environmental data vital to the nation's economic security, the safety of its citizens, and the understanding, protection, and management of our natural resources. The NOAA fleet of ships and aircraft is managed and operated by the Office of Marine and Aviation Operations (OMAO), an office comprising civilians, mariners, and officers of the NOAA Commissioned Officer Corps, one of the seven uniformed services of the United States. NOAA's roots trace back to 1807, when President Thomas Jefferson ordered the first comprehensive coastal surveys. Those early surveys ensured safe passage of ship-borne cargo for a young nation. As the needs of the nation have grown, so too have OMAO's responsibilities. Today, OMAO civilians and NOAA Corps officers operate, manage, and maintain NOAA's active fleet of 16 research and survey ships and nine specialized aircraft. Together, OMAO and the NOAA Corps support nearly all of NOAA's missions.

NOAA has the largest fleet of federal research and survey ships in the nation. The fleet ranges from large oceanographic ships capable of exploring and charting the world's deepest ocean, to smaller vessels responsible for surveying the shallow bays and inlets of the United States. The fleet supports a wide range of marine activities including fisheries surveys, nautical charting, and ocean and climate studies. Based throughout the continental United States, Alaska, and Hawaii, the ships operate in all regions of the nation and around the world.

NOAA's aircraft provide a wide range of airborne capabilities. Our highly specialized Lockheed WP-3D "Hurricane Hunter" aircraft are equipped with an unprecedented variety of scientific instrumentation, radars, and recording systems for both in situ and remote sensing measurements of the atmosphere, the Earth, and its environment. Equipped with both C-band weather radar and X-band tail Doppler radar systems, the WP-3Ds have the unique ability to conduct tropical cyclone research in addition to storm reconnaissance. Together with NOAA's Gulfstream IV-SP hurricane surveillance jet, these aircraft greatly improve our physical understanding of hurricanes and enhance the accuracy of tropical cyclone forecasts. NOAA's light aircraft also play a vital role in monitoring our environment. Our King Air, Commander and Twin Otter aircraft support marine mammal population studies, shoreline change assessments, oil spill investigations, and water resource/snowpack surveys for spring flood forecasts.

The NOAA fleet provides immediate response capabilities for unpredictable events. For example, in November 2014, our aircraft flew missions over upstate New York after the record snow falls of up to seven feet and conducted airborne Snow Water Equivalent (SWE) and soil moisture measurements. Airborne SWE measurements are used by NOAA's National Weather Service when issuing river and flood forecasts, water supply forecasts, and spring flood outlooks.

After Hurricane Sandy in 2012, NOAA ships Thomas Jefferson and Ferdinand R. Hassler conducted emergency bathometric surveys to locate possible submerged navigational hazards in the ports of New York and Virginia. These surveys enabled the ports to reopen quickly. Aerial images of storm-stricken regions, taken by NOAA aircraft, helped residents and emergency workers to quickly assess the condition of houses, bridges, and vital infrastructure. In 2010, the NOAA fleet and the NOAA Corps played a major role in the response to the BP Deepwater Horizon oil spill. NOAA's entire Atlantic fleet and over a quarter of the total strength of the NOAA Corps were deployed to the Gulf following the spill, developing mission plans and assisting response efforts.

While manned aircraft and sea-going vessels have been, and will continue to be, a primary source of environmental data, new technology will have a significant role to play in the future NOAA fleet. OMAO, in coordination with other NOAA offices and federal agencies, is evaluating and deploying remotely piloted underwater and aircraft systems that could significantly contribute to environmental observations. OMAO's ongoing challenge is to meet the growing demand for in situ scientific data while providing the highest level of service. To better serve the needs of the nation, NOAA is examining the composition of the fleet through an exhaustive and critical review of at-sea science and observation requirements. Our objective is to develop a clear, cost-efficient path forward to ensure that the NOAA fleet can continue to conduct at-sea surveys and research vital to fisheries management, updating nautical charts, responding to natural and manmade disasters, and understanding coastal and marine systems more fully. Meeting these requirements is essential to developing sustainable, science-based management and conservation plans that protect the health and resiliency of these resources over the long-term.

We continue our efforts to build a civilian and NOAA Corps officer work force that is uniquely qualified to gather critical environmental intelligence and be adaptive and responsive to a changing world and work to expand our partnerships with other federal agencies. For example, NOAA Corps officers are currently assigned to work in the Department of Defense, National Science Foundation, and the U.S. Senate among others where they lend their expertise and service. We also continue to strengthen our partnership with the U.S. Coast Guard. Our basic NOAA Corps officer training class is held at the U.S. Coast Guard Academy, where newly commissioned officers train alongside Coast Guard officer candidates, developing skills and professional relationships that will benefit both services, especially during challenging times. Active collaboration among the Federal family is critical to ensuring the long-term capability and success of the federal ocean infrastructure. Our partners' success is our success. The men and women of OMAO and the NOAA Corps provide environmental intelligence for a dynamic world as they serve our nation every day from the farthest seas to the highest skies.

NOAA Commissioned Officer Corps

– Honor, Respect, Commitment –

The NOAA Commissioned Officer Corps (NOAA Corps) is one of the nation's seven uniformed services and serve with the 'special trust and confidence' of the President. NOAA Corps officers are an integral part of the National Oceanic and Atmospheric Administration (NOAA), an agency of the U.S. Department of Commerce. With 321 officers, the NOAA Corps serves throughout the agency's line and staff offices to support nearly all of NOAA's programs and missions. The combination of commissioned service and scientific expertise makes these officers uniquely capable of leading some of NOAA's most important initiatives.

The NOAA Corps is part of NOAA's Office of Marine and Aviation Operations (OMAO) and traces its roots back to the former U.S. Coast and Geodetic Survey, which dates back to 1807 and President Thomas Jefferson. In 1970, NOAA was created to develop a coordinated approach to oceanographic and atmospheric research and subsequent legislation converted the commissioned officer corps to the NOAA Corps. The NOAA Corps today provides a cadre of professionals trained in engineering, earth sciences, oceanography, meteorology, fisheries science, and other related disciplines. Corps officers operate NOAA's ships, fly aircraft, manage research projects, conduct diving operations, and serve in staff positions throughout NOAA.

Benefits of the NOAA Corps to the Nation

The combination of commissioned service with scientific and operational expertise, allows the NOAA Corps to provide a unique and indispensable service to the nation. NOAA Corps officers enable NOAA to fulfill mission requirements, meet changing environmental concerns, take advantage of emerging technologies, and serve as environmental first responders. For example:

- In November 2014, our aircraft flew missions over upstate New York after the record snow falls of up to seven feet and conducted airborne Snow Water Equivalent (SWE) and soil moisture measurements. Airborne SWE measurements are used by NOAA's National Weather Service when issuing river and flood forecasts, water supply forecasts, and spring flood outlooks.

- After Hurricane Sandy in 2012, NOAA ships *Thomas Jefferson* and *Ferdinand R. Hassler* conducted emergency bathometric surveys to locate possible submerged navigational hazards in the ports of New York and Virginia. These surveys enabled the ports to reopen quickly. Aerial images of storm-stricken regions, taken by NOAA aircraft, helped residents and emergency workers to quickly assess the condition of houses, bridges, and vital infrastructure.

- After Hurricane Irene in 2011, the NOAA Ship *Ferdinand Hassler* and team completed 300 lineal nautical miles of survey work in less than 48 hours providing a Damage Assessment that enabled the U.S. Coast Guard to re-open ports and restore more than $5M per hour in maritime commerce less than three days after the storm.

- In 2010, the NOAA fleet and the NOAA Corps played a major role in the response to the BP Deepwater Horizon oil spill. NOAA's entire Atlantic fleet and over a quarter of the total strength of the NOAA Corps were deployed to the Gulf following the spill, developing mission plans and assisting response efforts.